LIFTOFF! SPACE EXPLORATION

OBSERVING EARTH

INVESTIGATING EARTH'S ATMOSPHERE

BY CHARLIE LIGHT

Gareth Stevens
PUBLISHING

Please visit our website, www.garethstevens.com. For a free color catalog of all our high-quality books, call toll free 1-800-542-2595 or fax 1-877-542-2596.

Cataloging-in-Publication Data

Names: Light, Charlie.
Title: Observing Earth: investigating Earth's atmosphere / Charlie Light.
Description: New York : Gareth Stevens Publishing, 2021. | Series: Liftoff! space exploration | Includes glossary and index.
Identifiers: ISBN 9781538258903 (pbk.) | ISBN 9781538258927 (library bound) | ISBN 9781538258910 (6 pack).
Subjects: LCSH: Atmosphere–Juvenile literature. | Earth (Planet)–Juvenile literature.
Classification: LCC QC863.5 L54 2021 | DDC 551.51–dc23

First Edition

Published in 2021 by
Gareth Stevens Publishing
111 East 14th Street, Suite 349
New York, NY 10003

Designer: Katelyn E. Reynolds
Editor: Abby Badach Doyle

Photo credits: Cover, p. 1 xtock/Shutterstock.com; cover, pp. 1–32 (series background) David M. Schrader/Shutterstock.com; p. 5 BigMouse/Shutterstock.com; p. 7 Enrique Díaz / 7cero/Moment/Getty Images; p. 8 Library of Congress/Getty Images; pp. 9, 13 ullstein bild/ullstein bild via Getty Images; p. 11 sansubba/E+/Getty Images; p. 12 Hulton Archive/Getty Images; p. 15 Dneutral Han/Moment/Getty Images; pp. 16, 23 (inset) JSC/NASA; p. 17 Alex Gerst/JSC/NASA; p. 19 MSFC/NASA; p. 21 GSFC/NASA Goddard; p. 23 (main) ansonmiao/E+/Getty Images; p. 25 Sovfoto/Universal Images Group via Getty Images; p. 27 Fine Art Images/Heritage Images/Getty Images; p. 29 JPL/wNASA/GSFC.

Printed in the United States of America

CONTENTS

Exploring Earth's Atmosphere 4
Take Off to the Troposphere 6
Shooting for the Stratosphere 10
Marvels in the Mesosphere 14
Traveling to the Thermosphere 18
Into the Ionosphere 20
Meet the Magnetosphere 22
Excitement in the Exosphere 24
Explore or War? 26
Exiting Earth's Atmosphere 28
Glossary 30
For More Information 31
Index 32

WORDS IN THE GLOSSARY APPEAR IN **BOLD** TYPE THE FIRST TIME THEY ARE USED IN THE TEXT.

EXPLORING EARTH'S ATMOSPHERE

Did you know that you're living inside a bubble? You can't see it, but there's a bubble wrapped around our planet. It's made of gases that you can't see. It's called the atmosphere, and it makes life on Earth possible! You already know the atmosphere by another name—air. The atmosphere is the air you breathe. It has oxygen that you breathe in and carbon dioxide that you breathe out.

Earth's atmosphere has other jobs too. It **absorbs** heat from the sun to keep Earth at a steady, warm temperature. Without it, we'd be too cold at night and too hot during the day. At the same time, the atmosphere protects us from **radiation** from the sun. It also keeps most space objects from reaching Earth's surface.

INTERSTELLAR INFORMATION

THE ATMOSPHERE IS HELD CLOSE TO OUR PLANET BY A FORCE CALLED GRAVITY. THIS IS A PULLING FORCE. IT PULLS THE GASES OF THE ATMOSPHERE TOWARD EARTH.

THE FIVE LAYERS OF EARTH'S ATMOSPHERE

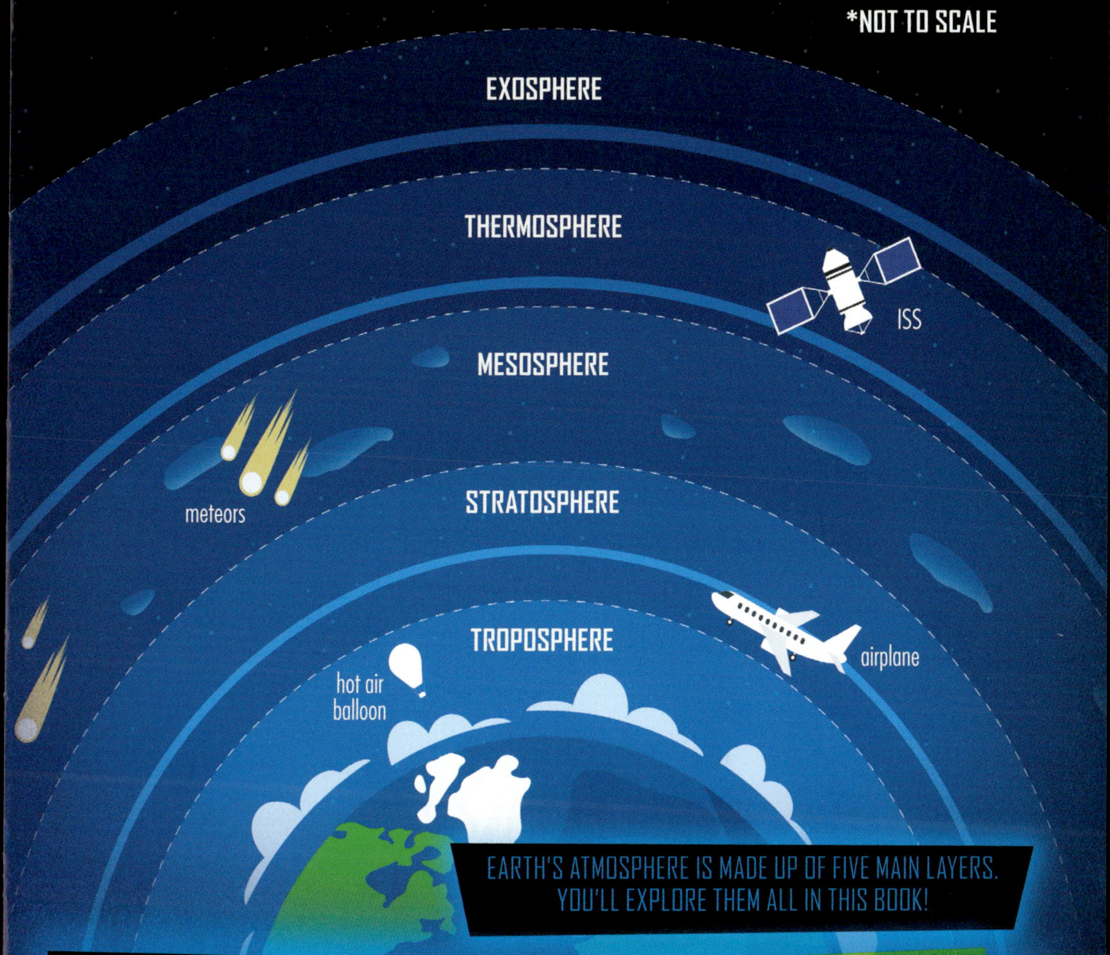

EARTH'S ATMOSPHERE IS MADE UP OF FIVE MAIN LAYERS. YOU'LL EXPLORE THEM ALL IN THIS BOOK!

AERODYNAMICS

The study of how objects move through gases is called aerodynamics (air-oh-dy-NAM-iks). This branch of science let humans explore Earth's atmosphere and beyond. Aerodynamics looks at forces that make it possible to fly. A force is a push or a pull on an object that makes it move. Pulling brings an object closer to you. Pushing moves an object farther away from you. Forces happen when two objects interact. When the interaction stops, the force stops too.

TAKE OFF TO THE TROPOSPHERE

The troposphere is the lowest layer of Earth's atmosphere. You're already in it! This is the layer of the atmosphere where humans live and breathe, and where birds fly. It reaches 5 to 9 miles (8 to 14.5 km) above the planet's surface. Most human pollution collects here too. The troposphere is also the layer where the water cycle takes place. That's because 99 percent of all water vapor in the atmosphere is found here. This layer is where most clouds form and nearly all weather occurs. Changes in temperature, humidity, precipitation, air pressure, wind, and clouds are all part of weather.

Before humans could make it to space, they had to learn to fly. Early attempts at human flight happened in the troposphere.

INTERSTELLAR INFORMATION

THE TROPOSPHERE IS NAMED FOR THE CONSTANT CHANGES OF WEATHER THAT HAPPEN THERE. "TROPOS" MEANS CHANGE.

MOST LIGHTNING OCCURS IN THE TROPOSPHERE, BUT SOMETIMES THUNDERSTORMS AND OTHER STORMS TAKE PLACE AS HIGH AS THE STRATOSPHERE!

WHO TURNED UP THE HEAT?

There is a tiny amount of carbon dioxide in the troposphere. The amount has almost doubled since 1900. This is very dangerous to life on our planet because carbon dioxide is a greenhouse gas. Greenhouse gases in Earth's atmosphere stop heat from escaping our planet and releasing into space. This makes Earth's temperature rise. Extra greenhouse gases caused by humans are making Earth warmer. This is known as climate change.

Humans have been trying to fly for hundreds of years. The first invention that let humans fly was the hot air balloon. It was created by the Montgolfier brothers, Joseph Michel and Jacques Etienne. The first human flight in their balloon happened in 1783.

Over the next century, several inventors worked on gliders. These machines were supposed to allow a single person to fly. Many had bird-like wings and a tail for steering. Otto Lilienthal was the first person to make a glider that could fly a person over long distances.

Orville and Wilbur Wright studied Lilienthal's work. The Wright brothers were the first people to invent a working, powered plane. Their first successful airplane flight was in 1903.

INTERSTELLAR INFORMATION

A SHEEP, A ROOSTER, AND A DUCK WERE THE FIRST PASSENGERS TO FLY IN THE MONTGOLFIERS' HOT AIR BALLOON! THEY FLEW UP 6,000 FEET (1,829 M) AND TRAVELED MORE THAN 1 MILE (1.6 KM).

WRIGHT BROTHERS' FIRST FLIGHT

OTTO LILIENTHAL FLEW ON HIS GLIDERS AS MANY AS 2,000 TIMES. HE PASSED AWAY AFTER CRASHING A GLIDER IN 1896.

PUTTING THE "LIFT" IN LIFTOFF

Lift is the force that holds objects up in the air. When an airplane moves through the air, it causes lift. Most lift comes from the plane's wings. The shape of the wings makes the air move faster over the top of the wings. As the air moves faster, the pressure of the air lessens. There is more pressure underneath the wings than on top of the wings. This difference in air pressure creates lift.

SHOOTING FOR THE STRATOSPHERE

The stratosphere reaches from about 7.5 miles (12 km) to 31 miles (50 km) above Earth's surface. Humans can't survive here without protection. The atmospheric pressure is too low.

Most airplanes fly in the stratosphere because there is almost no weather here. In this layer, the coldest air is at the bottom and the warmest air is at the top. That's the opposite of how heat works in the troposphere.

"Strat" means layer. The stratosphere has a layer of its own—the ozone layer. Ozone is a special type of oxygen **molecule**. The ozone layer absorbs radiation from the sun. Without the ozone layer's protection, the sun's radiation would hurt or even kill most life on Earth.

INTERSTELLAR INFORMATION

THE LOWEST AIR PRESSURE HUMANS CAN SURVIVE IN IS AT ABOUT 5 MILES (8 KM) ABOVE **SEA LEVEL**. THE TOP OF MT. EVEREST IS 5.5 MILES (9 KM) ABOVE SEA LEVEL.

THE "DEATH ZONE" ATOP MT. EVEREST STARTS AT 26,247 FEET (8,000 M). HERE, YOUR BODY SLOWLY DIES FROM THE LOW PRESSURE. THIS IS STILL 2 MILES (3.2 KM) BELOW THE STRATOSPHERE.

ATMOSPHERIC PRESSURE

Atmospheric pressure measures how much the atmosphere is pushing down on Earth. You can also look at it as how heavy the atmosphere is. Earth's atmosphere is heaviest at sea level. The higher above sea level you go, the lighter our atmosphere becomes. Humans need an atmosphere that is heavy enough to help us breathe. Atmospheric pressure pushes the air into our lungs and pushes oxygen through our blood. Too little pressure—or too much—will kill us.

Swiss scientist Auguste Piccard invented the first balloon that could safely reach the stratosphere. He built a gondola, or cabin, under the balloon. It was pressurized. This meant the inside of the cabin stayed at an atmospheric pressure comfortable to humans. It also had a system that pumped in oxygen and took out carbon dioxide. This let passengers breathe safely.

On May 27, 1931, Piccard and his helper, Dr. Paul Kipfer, flew this balloon to the greatest height yet. They went up 51,577 feet (15,720 m), or nearly 10 miles (16 km). That's into the stratosphere!

Piccard's pressurized gondola became an example for future inventions. Now scientists knew they would need to build pressurized aircraft to explore higher levels of the atmosphere—and even outer space.

INTERSTELLAR INFORMATION

THE FIRST WOMAN TO REACH THE STRATOSPHERE WAS SCIENTIST DR. JEANNETTE PICCARD. SHE WAS MARRIED TO AUGUSTE PICCARD'S TWIN BROTHER. SHE FLEW AUGUSTE PICCARD'S BALLOON 57,579 FEET (17,550 M) ABOVE SEA LEVEL.

PICCARD'S BALLOON WAS FILLED WITH HYDROGEN GAS. HYDROGEN IS LIGHTER THAN THE TWO MOST COMMON GASES IN THE AIR—NITROGEN AND OXYGEN. THIS IS WHY A HYDROGEN BALLOON FLOATS.

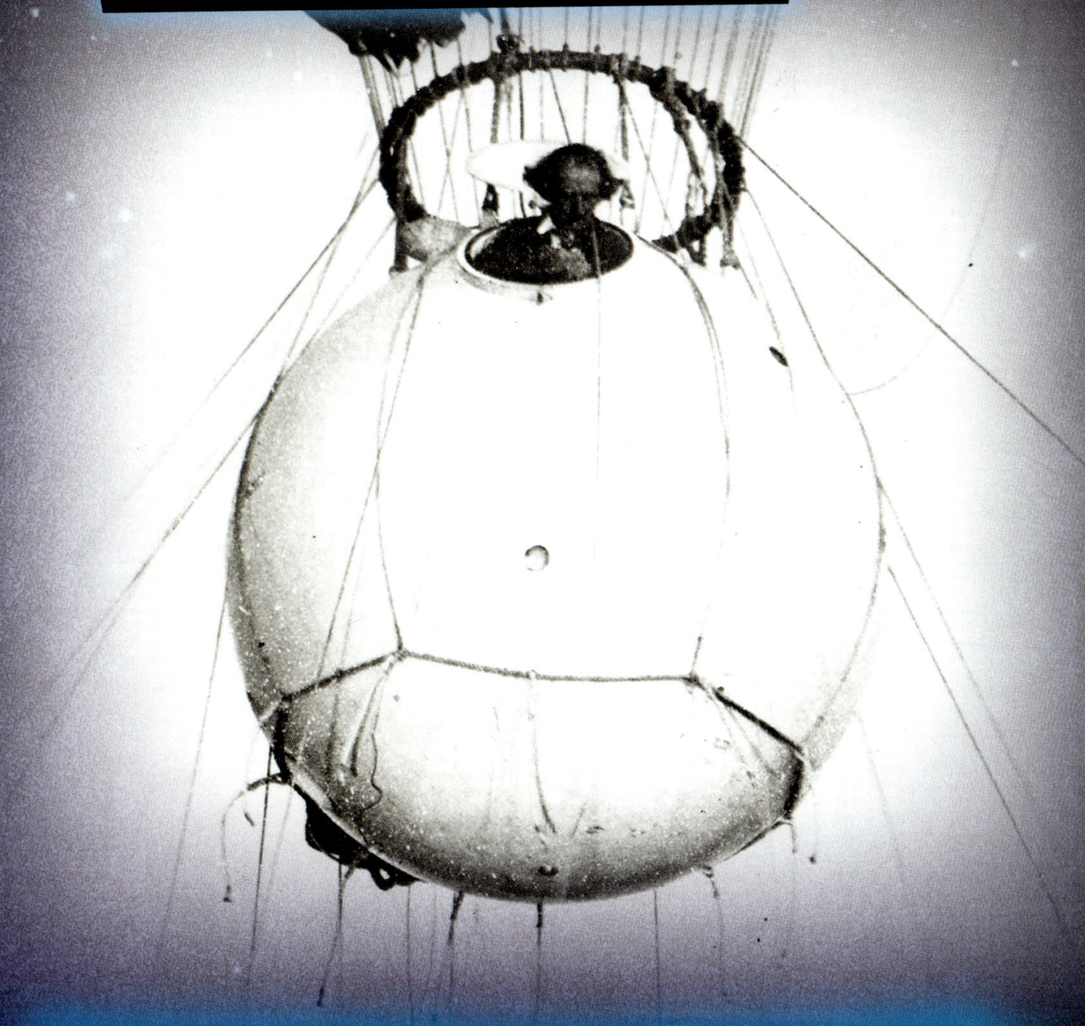

THE RACE BEGINS

After Piccard's flight in 1931, the United States and the former Union of Soviet Socialist Republics (USSR) started to build their own balloons. This was reported to be a friendly race. The USSR, which is called Russia today, beat Piccard's record first using a larger gondola with better technology. It flew more than 11 miles (18 km) high in September 1933. In November of that year, a new U.S. balloon flew even higher. It beat the USSR's record by 542 feet (165 m).

MARVELS IN THE MESOSPHERE

The mesosphere is where the atmospheric layers start to get colder. It stretches from about 31 to 50 miles (50 to 80 km) above Earth's surface. The top of the mesosphere is -120°F (-85°C). It's the coldest part of the atmosphere and colder than any place on Earth.

The mesosphere acts like a forcefield, or shield, around the planet. Shooting stars occur in the mesosphere. Shooting stars are actually meteors. A meteor is a hunk of metal or rock that enters Earth's atmosphere moving at a high speed. It moves so fast through the mesosphere that it burns up. This is how the mesosphere keeps Earth safe from getting hit.

INTERSTELLAR INFORMATION

BEFORE A METEOR ENTERS EARTH'S ATMOSPHERE, IT'S CALLED A METEOROID. INSIDE THE ATMOSPHERE, IT'S CALLED A METEOR. IF A METEOR REACHES EARTH'S SURFACE, IT BECOMES A METEORITE.

IN A METEOR SHOWER, MANY METEORS BURN UP IN THE MESOSPHERE AT ONCE. THIS HAPPENS WHEN EARTH MOVES THROUGH A PATCH OF **DEBRIS**.

FRICTION

Why do meteors burn up in the mesosphere if it's so cold there? The answer is friction. Friction is a force that happens when two objects rub against each other. Friction slows down motion. It also creates heat. Meteors that reach the mesosphere are moving very fast. They rub against the gases in this layer as they move. This causes friction, which heats the meteors up. Nearly all meteors burn up before they can reach Earth.

The mesosphere is a mystery in the middle of our atmosphere. "Meso" means middle. It's too high for regular airplanes to reach. Balloons can only reach the very bottom of this layer. It's also too low for **satellites** to orbit. Scientists have been able to find some information about this layer by using sounding rockets. These are rockets used for research.

Sounding rockets carry equipment. They have two main parts—a rocket motor and a payload carrying equipment. The motor pushes the sounding rocket into the atmosphere. When it's out of fuel, it breaks off from the payload and falls back to Earth. The payload keeps moving up into space. It records data with its equipment. Then, it falls back to Earth with a parachute.

INTERSTELLAR INFORMATION

SINCE METEORS TURN TO VAPOR IN THE MESOSPHERE, THIS LAYER HAS BEEN FOUND TO HAVE HIGHER AMOUNTS OF IRON AND METAL THAN OTHER LAYERS OF OUR ATMOSPHERE.

MESOSPHERIC CLOUDS AT SUNRISE, AS VIEWED FROM SPACE

THIS IMAGE WAS TAKEN FROM THE INTERNATIONAL SPACE STATION (ISS) IN THE THERMOSPHERE, ABOVE THE MESOSPHERE. IT SHOWS MESOSPHERIC CLOUDS FORMING OVER CANADA.

SPRITES AND ELVES IN SPACE?

The mesosphere is much drier than the troposphere. However, sometimes clouds still form here. They are called mesospheric polar clouds because they form above Earth's poles. These types of clouds form much higher than any other type of cloud. There are also strange bursts of electricity in the mesosphere. These bursts are like lightning, but they happen far above where most thunderstorms form. Scientists call these bursts sprites and elves.

TRAVELING TO THE THERMOSPHERE

The thermosphere is a warm part of the atmosphere. "Thermo" means heat. It can get up to 4,500°F (2,482°C) here! This layer is between 50 and 440 miles (80 and 700 km) above Earth's surface.

Even though the thermosphere is hot, it would feel freezing to you. Humans need gas molecules to transfer heat to our bodies. There aren't enough gas **particles** in the thermosphere to transfer heat to you. There aren't enough for **sound waves** to travel here, either. That means there's no sound in the thermosphere.

Even though the thermosphere is much higher than where humans live, there are many human-made objects that exist here. The ISS is in this layer. Many low-flying satellites orbit Earth in the thermosphere too.

INTERSTELLAR INFORMATION

THE SUN GIVES OFF NEVER-ENDING WAVES OF ENERGY PARTICLES CALLED SOLAR WINDS. IT ALSO HAS SOLAR FLARES, OR POWERFUL BURSTS OF ENERGY FROM A SMALL PART OF THE SUN'S SURFACE.

THE ISS IS A LABORATORY THAT ORBITS EARTH TO STUDY SPACE. IT WAS BUILT BY 15 NATIONS WORKING TOGETHER. THERE HAVE BEEN CREWS ON BOARD SINCE 2000.

STRETCHING WITH THE SUN

The thermosphere absorbs particles with lots of energy coming off of the sun. These particles are what heat up the thermosphere. That's why this layer is so hot. The sun has different levels of activity. The more active the sun is, the more particles it gives off—and the hotter the thermosphere gets. When the sun is very active, the top of the thermosphere actually puffs up! This layer stretches higher into space as it gets hotter.

INTO THE IONOSPHERE

The ionosphere is a special part of Earth's atmosphere that is different from the others. It's actually a group of smaller areas spread out through the mesosphere, thermosphere, and exosphere. All these layers have areas, or pockets, of particles called ions. Ions are particles with an electric charge. Ions have an uneven number of electrons and protons. The ionosphere is the name for these pockets of ions.

The ionosphere is formed by energy from the sun hitting the gases that make our atmosphere. When solar energy hits a gas particle, it sometimes knocks out the particle's electrons. The uneven particle becomes an ion. The free electron floats around on its own.

INTERSTELLAR INFORMATION

THE IONOSPHERE GROWS AND SHRINKS IN SIZE. ITS SIZE DEPENDS ON THE AMOUNT OF ENERGY IT TAKES IN FROM THE SUN.

SOLAR FLARES, LIKE THIS ONE, GIVE OFF HUGE AMOUNTS OF SOLAR ENERGY. THE BIGGEST ONES CAN EVEN INTERRUPT RADIO COMMUNICATIONS AND THE POWER GRID ON EARTH.

PROTONS, NEUTRONS, AND ELECTRONS

To understand the ionosphere, it helps to understand the tiny charged particles that exist there. Particles are made up of even smaller parts called neutrons, electrons, and protons. A neutron has no charge. Electrons have a negative charge. Protons have a positive charge. Ions with more electrons than protons have a negative charge. Ions with more protons than electrons have a positive charge. Particles with the same number of electrons and protons have no charge.

MEET THE MAGNETOSPHERE

The ionosphere is where auroras form. You may know these colorful sights as the northern and southern lights. Auroras happen when the ionosphere works with another invisible bubble surrounding the planet—the magnetosphere. This is a **magnetic field** around Earth. Scientists think it's caused by metals in Earth's core, or center.

The magnetosphere isn't part of Earth's atmosphere. However, it does help protect it! The magnetosphere acts like a shield around the atmosphere and Earth itself. It protects us from solar winds and other powerful energy particles. The magnetosphere absorbs this energy, then it shoots the energy back out. The magnetosphere is shaped by the solar winds pushing into it. Its shape changes with these winds all the time.

INTERSTELLAR INFORMATION

THE NORTHERN LIGHTS ARE CALLED THE AURORA BOREALIS. THE SOUTHERN LIGHTS ARE CALLED THE AURORA AUSTRALIS.

DIFFERENT GASES CREATE THE COLORS WE SEE IN AURORAS. OXYGEN MAKES REDS AND GREENS. NITROGEN MAKES BLUES AND PURPLES.

AURORA BOREALIS AS SEEN FROM THE INTERNATIONAL SPACE STATION ▶

WHY DO AURORAS HAPPEN?

Auroras happen when the sun has storms. Solar storms send extra powerful solar winds over Earth. Some of the solar energy particles bounce off the magnetosphere and move away from Earth. However, some particles move down lines in the magnetosphere. These lines go down to the Earth's North and South poles. There, the energy particles mix with gases in the ionosphere. This mixing makes auroras.

EXCITEMENT IN THE EXOSPHERE

The outermost layer of Earth's atmosphere is known as the exosphere. "Exo" means "outside." This layer exists between 440 to 6,200 miles (700 to 10,000 km) above the planet's surface. It's very cold here.

The exosphere is very big. It's about 6,200 miles (10,000 km) thick. That's almost as wide as Earth itself! It's much thicker than any other layer of the atmosphere. Here, there are more satellites orbiting Earth than in the thermosphere. In fact, most satellites that orbit the Earth are in this layer. The air is so thin in the exosphere that it's almost like outer space. This layer doesn't have a clear "roof" to show where it ends. The exosphere just blends into outer space.

INTERSTELLAR INFORMATION

SPUTNIK I WAS THE FIRST SATELLITE MADE BY HUMANS TO ORBIT EARTH. IT WAS SLIGHTLY BIGGER THAN A BASKETBALL AND WEIGHED 183 POUNDS (83 KG).

SPUTNIK 1 WAS LAUNCHED BY THE USSR ON OCTOBER 4, 1957. IT REACHED THE EXOSPHERE AT ITS **MAXIMUM** ALTITUDE OF 583 MILES (939 KM).

WHERE DOES IT END?

Scientists aren't sure where Earth's atmosphere ends and space begins. Some think the exosphere is actually part of outer space, not our atmosphere. Others think Earth's atmosphere stretches farther than the exosphere—maybe even beyond the moon. One way of measuring where "space" starts is the Kármán Line. This is about 62 miles (100 km) above Earth's surface. This is where spacecraft start to escape the pull of Earth's gravity. The Kármán Line is in the thermosphere.

EXPLORE OR WAR?

Sputnik 1's launch started a time in history known as the Space Race. The United States and the USSR were racing to explore the higher layers of our atmosphere and beyond. This time, it wasn't a friendly race. It was part of the Cold War. The Soviets were ahead for much of the race. They sent the Luna 1 spacecraft outside Earth's atmosphere in 1959. Luna 1 passed the moon and became the first human-made object to orbit the sun.

Then, the USSR's *Vostok 1* mission sent the first human into space. On April 12, 1961, **cosmonaut** Yuri Gagarin made one orbit around Earth. It took less than an hour and a half. He reached the thermosphere at a top height of 203 miles (327 km).

INTERSTELLAR INFORMATION

AFTER SPUTNIK I WAS LAUNCHED, THE U.S. CONGRESS PASSED THE "SPACE ACT" IN 1958. THIS CREATED THE NATIONAL AERONAUTICS AND SPACE ADMINISTRATION, BETTER KNOWN AS NASA.

LAIKA THE DOG WAS THE FIRST ANIMAL TO ORBIT EARTH. SHE WAS LAUNCHED ON SOVIET RUSSIA'S SPUTNIK 2 SATELLITE ON NOVEMBER 3, 1957.

THE COLD WAR

The Cold War was a time of **tension** between the United States and the USSR. It lasted from 1946 to 1991, when the USSR **collapsed**. Both nations were trying to prove they were more powerful than the other. To show this, they raced to invent new technology. This included the Space Race. It also included an arms race to see which country could build the most and the strongest nuclear weapons.

EXITING EARTH'S ATMOSPHERE

Our missions to explore Earth's atmosphere were an important first step for space exploration that followed. In December 1968, the U.S. sent the first humans out of Earth's orbit in the *Apollo 8* mission. In July 1969, astronaut Neil Armstrong became the first human to walk on the moon in the United States' *Apollo 11* mission. The Space Race with the USSR ended by 1972.

Today's technology lets us study other planets' atmospheres! Having this as a comparison helps scientists understand Earth's atmosphere better. NASA's MAVEN and Curiosity missions found that Mars probably had an atmosphere like Earth's billions of years ago. NASA has even found planets that might be like Earth by using the Kepler Telescope. Maybe you'll be the first human to walk on another planet with an Earth-like atmosphere!

INTERSTELLAR INFORMATION

THE OUTERMOST PART OF EARTH'S EXOSPHERE IS KNOWN AS THE GEOCORONA. PARTS OF IT HAVE BEEN FOUND PAST THE MOON'S ORBIT, ABOUT 400,000 MILES (630,000 KM) AWAY!

SCIENTISTS THINK THAT STRONG SOLAR WINDS WORE DOWN MARS'S ATMOSPHERE OVER TIME. THIS IS WHY THE PLANET IS DRY AND BARREN TODAY.

PEACE IN SPACE

There's still much to explore in our own atmosphere. One place we can do this is the ISS. Although exploring space began as a race, today 15 nations work together on the ISS. It's run by five space programs—NASA (United States), Roscosmos (Russia), ESA (Europe), JAXA (Japan), and CSA (Canada). More than 200 people from 19 countries have been to the ISS. As many as six spaceships can be connected to the station at one time.

GLOSSARY

absorb: to take in

collapse: to fail or stop working

cosmonaut: Soviet or Russian astronaut

debris: the remains of something that has been broken

magnetic field: the area around a magnet where its pull is felt. Earth has a magnetic field too.

maximum: the highest number or amount that is possible or allowed

molecule: a very small piece of matter

particle: a very small piece of something

radiation: waves of energy

satellite: an object that circles Earth. Also, an object that circles Earth in order to collect and send information or aid in communication.

sea level: the average level of the ocean

sound waves: a wave carries sound to your ear through the air

tension: a state in which countries disagree and feel anger toward each other

FOR MORE INFORMATION

BOOKS

Borngraber, Elizabeth. *The Layers of Earth's Atmosphere.* New York, NY: PowerKids Press, 2018.

Shofner, Melissa Rae. *Earth's Atmosphere.* New York, NY: PowerKids Press, 2018.

Space Travel. New York, NY: DK Children, 2019.

WEBSITES

DK Findout: The Space Race
www.dkfindout.com/us/space/space-race/
Learn more about the Space Race at this interactive site.

Kids' Crossing: Stuff in the Sky
eo.ucar.edu/kids/sky/air1.htm
The National Center for Atmospheric Research has this site just for kids. Learn fun facts, play games, and more.

NASA Space Place
spaceplace.nasa.gov/menu/earth/
Explore Earth and its atmosphere at NASA's site for kids.

INDEX

aerodynamics 5

Apollo missions 28

atmospheric pressure 10, 11, 12

carbon dioxide 4, 7, 12

climate change 7

clouds 6, 16, 17

electricity 17

electrons 20, 21

force 4, 5, 9, 15

friction 15

human flight 6, 8

International Space Station (ISS) 17, 18, 19, 29

Kármán Line 25

Kepler Telescope 28

lift 9

meteor 14, 15, 16

NASA 26, 28, 29

ozone layer 10

protons 20, 21

radiation 4, 10

satellites 16, 18, 24, 27

sea level 10, 11, 12

solar winds 18, 22, 23, 29

sounding rockets 16

sound waves 18

Space Race 26, 27, 28

Sputnik 1 24, 25, 26, 27

water cycle 6